YOUR KNOWLEDGE HAS VALUE

- We will publish your bachelor's and master's thesis, essays and papers

- Your own eBook and book - sold worldwide in all relevant shops

- Earn money with each sale

Upload your text at www.GRIN.com and publish for free

Bibliographic information published by the German National Library:

The German National Library lists this publication in the National Bibliography;
detailed bibliographic data are available on the Internet at http://dnb.dnb.de .

Imprint:

Copyright © 2016 GRIN Verlag, Open Publishing GmbH
Print and binding: Books on Demand GmbH, Norderstedt Germany
ISBN: 9783656988076

This book at GRIN:

http://www.grin.com/en/e-book/323503/estimating-electron-density-of-the-ionos-
phere-from-radio-occultation-tec

Bizuayehu Adisie Beyene

Estimating Electron Density of the Ionosphere from Radio Occultation TEC Data using Abel Inversion

GRIN Publishing

Estimating Electron Density of the Ionosphere from Radio Occultation TEC Data using Abel Inversion

Bizuayehu Adisie Beyene

Physics Department, Arbaminch University, Arbaminch, 21, Ethiopia

---

Abstract Ionospheric electron density affects trans-ionospheric radio waves. Estimating electron density, therefore, is vital to mitigate the effect of ionosphere on radio waves. As a result, different methods have been applied to estimate ionospheric electron density from integrated measurements of ground-based GPS (Global Positioning System) receivers recently installed in East-Africa. To augment such efforts, ionospheric electron density has been estimated from Radio Occultation (RO) total electron content (TEC) data measured by LEO (Low Earth Orbit) satellites. Abel inversion onion peeling algorithm then is used to retrieve electron density profiles from the calibrated TEC data. The reconstructed electron density profiles have been validated by running the standard ionospheric empirical model (International Reference Ionosphere, IRI-2007) at time and locations that RO electron density profiles are obtained. The results show well agreement between the Abel inversion and IRI-2007 electron density profiles, for both bottom and topside ionosphere. Also, F_2 peak density heights from Abel inversion and IRI-2007 have shown well agreement. However, significant discrepancy has been detected around the height of the F_2 peak density.

Keywords: Abel inversion, electron density, radio occultation, total electron content

Contents

1. Introduction

Ionosphere is the ionized part of the Earth's upper atmosphere which contains free electrons and positive ions. It is formed when solar radiation strikes the upper atmosphere in such a way that the incoming radiation with sufficient energy takes away electrons from the neutral atom and hence ionized medium is formed. The electrons formed as a result of the ionization process in general and Total Electron Content (TEC) in particular are the main characterizing features of the ionosphere. TEC is defined as the integration of the free electron distribution in a $1m^2$ column along the signal path from the satellite to the receiver. The variation of these parameters describes the variations of the ionosphere. Some of the determinants of ionosphere variation are solar radiation variation, thermospheric wind, deposition of energetic particles at higher latitudes and variation in neutral composition of the atmosphere [2].

The variation of the ionosphere affects trans-ionospheric radio wave users, for example, communication and GPS (Global Positioning System) applications such as positioning and navigation. A proper way of understanding the ionospheric variability can help us to use the appropriate signals that are used for communication purposes.

Ionospheric effects on the radio waves can be mitigated in many ways. Linear combinations of signals with different frequencies can be used for mitigating the effect of ionosphere on global navigation satellite system (GNSS) measurements. On the other hand, the integration of free electrons that are on the radio path, TEC estimation can be used to reduce the effects of the dispersive medium (ionosphere) on communication and navigation technology [7]. It is also used to study about the properties of the ionospheric regions for ionospheric research. Ionospheric TEC can be estimated from ground and space-based GPS measurements. The space-based GPS receivers are carried by the LEO (Low Earth Orbit) satellites. The measurement system using GPS receivers carried by LEO satellites is called radio occultation (RO). RO is a remote sensing technique which is used to study the atmosphere of planets from an appropriate transmitter (GPS) - receiver (LEO) combination. The receivers carried by the LEO track signals from the constellation of GPS satellites at $20,000km$ altitude.

Incoherent Scatter Radar (ISR) techniques and Ionosonds can be used for ionospheric measurements. ISR is based on sending radio signals to probe the ionosphere and helps us to get information from the reflected echo. When a radio wave is sent in to the ionosphere, free electrons can scatter the waves. The strength of the echo received from the ionosphere can be used to estimate the number of electrons in the scattering volume; hence electron density can be determined this way. Ionosonde can be used to estimate ionospheric electron density and is based on sending radio signals to the ionospheric region and studying the properties of the reflected signals. From the heights of the reflective ionospheric region, vertical profiles of electron density can be found. From ground-based GPS measurement of TEC data, electron density profile can be estimated using Tomographic inversion method (Endawoke et al., 2007). Tomographic inversion is based on dividing the ionosphere into separate boxes (voxels), and estimating one value of electron density for each box. Also, Abel inversion can be used to retrieve ionospheric electron density profile from space based TEC measurement data. Starting from the upper part of the ionospheric layer, we ingest TEC data in to the Abel inversion algorithm and go inward until the integrated sum of free electrons on the radio path is computed. This algorithm is called onion peeling algorithm as it seems peeling an onion (Nava, 2012). As compared to Abel inversion method, Tomographic inversion takes relatively longer computational time and there is sacristy of data in ground based measurement as receivers can't be deployed everywhere (Skone, 2010).

While a lot of work is done on estimating TEC for the East African ionosphere from ground based GPS measurement (Melessew et al., 2013), electron density retrieval from radio occultation TEC data is not yet done here in our country. Therefore, in this paper, electron density of the ionosphere is going to be estimated from radio occultation COSMIC TEC data using Abel inversion method.

Objective:

• Retrieving electron density profile from radio occultation TEC data using onion peeling algorithm.

• Comparing radio occultation electron density profiles with electron density profiles estimated by standard model (international reference ionosphere, IRI).

The main contribution of this work is the Matlab code developed to implement the Abel inversion onion peeling algorithm.

Also, the outputs of the Abel inversion algorithm can be used to validate others work that will be carried out in the future. Moreover, it can be utilized as springboard for further study related to radio occultation observations.

2. Methodology

As ionosphere is a dispersive medium, radio waves propagating through it may take relatively longer times to arrive at the receiver end than the normal straight line trajectory. This ionospheric delay is directly proportional to the slant total electron content ($sTEC$) along the radio path and inversely related to the radio frequency. As the 6-COSMIC satellites rotate in a low earth orbit, they acquire radio signals from GPS satellites which are at $20,000km$ altitude so;

ionospheric effect is contained in the radio signals that are measured in the receiver. The basic observable for each occultation is the phase change between the transmitter and the receiver as the signal propagates through the ionosphere and the neutral atmosphere. $sTEC$ can then be extracted from phase observations of the radio occultation measurements on the LEO receiver. This $sTEC$ data is then feed to the Abel inversion onion peeling algorithm to get electron density profiles of the ionosphere.

The methodology of this study is based on Abel inversion of $sTEC$ data under the assumption of spherical symmetry. Ionospheric electron density profile retrieval using Abel inversion is based on spherical symmetry of electron density in the occultation plane. Spherical symmetry means that the retrieved electron density depends only on height. The LEO satellites are rotating in nearly circular orbits and radio signals propagate in a straight line from the transmitter to the receiver. First order estimation of electron density is done up to the LEO satellites altitude (Schreiner et al., 2009). Abel inversion algorithm for electron density of the ionosphere to be:

$$N(p) = \frac{1}{\pi} \int_{r=p_0}^{r=p} \frac{\frac{dTEC}{dp}}{\sqrt{p^2 - r^2}} dp \qquad (2.1)$$

Where p is the impact parameter, r is the radial distance and N is ionospheric electron density and TEC is the total electron content on the radio path..

For practical applications, the above Abelian inversion for the electron density of the ionosphere (Equation 2.1) can be formulated in another way by using the discretized value of the TEC along the signal path.

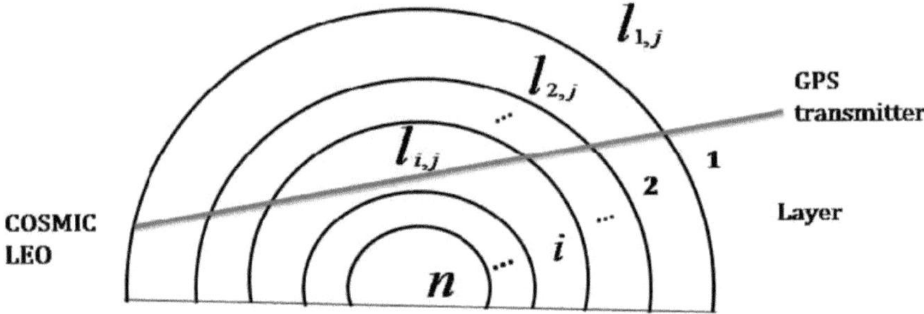

Figure 2.1: RO geometry and divided ionospheric layers (Adapted from Ouyang, 2008b).

The assumption is that electron density in each layer has a uniform distribution.

Having the COSMIC height to be approximately 800km, we can divide the ionosphere in to n layers as shown in figure 2.1 above. Also, it is assumed that the electron density is nearly zero below 50km and above 1000km since there is no significant number of free electrons in these regions.

The TEC along the radio path then is given as (particularly for each layer i i.e. for i^{th} layer and j^{th} ray path (Figure 2.1));

$$TEC_j = 2.\sum_{j=1}^{i} N_j l_{i,j} \qquad (2.2)$$

Which is the same as

$$TEC_j = 2.\sum_{j=1}^{i-1} N_j l_{i,j} + 2. N_i l_{i,i} \qquad (2.3)$$

Solving for N_i from the above equation, we get;

$$N_i = \frac{TEC_j - 2.\sum_{j=1}^{i-1} N_j l_{i,j}}{2. l_{i,i}} \qquad (2.4)$$

Where, $l_{i,i}$ is the path length between the ionospheric layers. Equation 2.4 can be solved iteratively starting from the outer most layer $(i = 1)$ to the bottom layer $(i = n)$ like peeling an onion (Ouyang, 2008a and the references therein).

To determine ionospheric electron density using onion peeling algorithm, the length of the ray path between layers, $l_{i,i}$, must at first be known. Under the spherical symmetry assumption, where the LEO orbit is nearly circular with radius equal to the impact parameter (p), we can have the following occultation geometry.

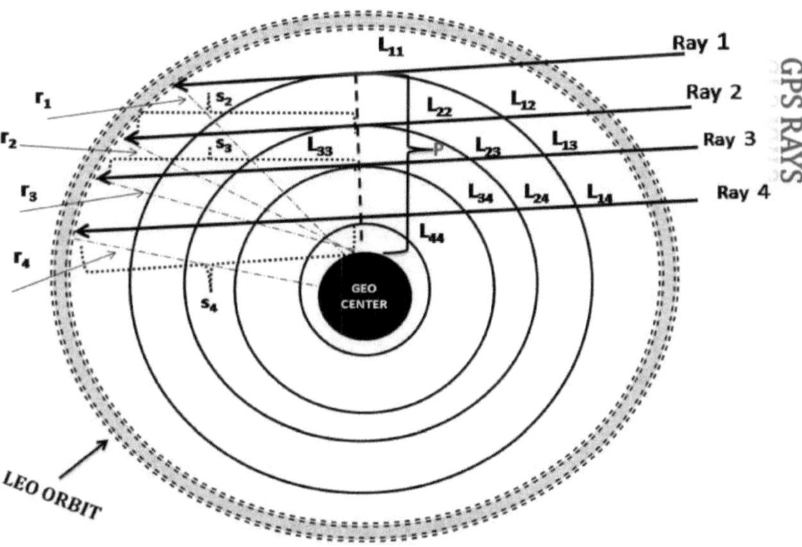

Figure 2.2: RO geometry and discretized ionosphere (Adapted from Nava, 2012).

From figure 2.2, it can be seen that p is the impact parameter and r_1, r_2, r_3 and r_4 are the corresponding LEO heights during the occultation event. As each GPS ray path and the impact parameter are perpendicular to each other, we can have the following relations for each of the four rays:

$$\frac{l_{1,1}^2}{2} + p_1^2 = r_1^2 \qquad (2.5)$$

Therefore, for the first ray in figure 2.2, the length of the first ray path is given by;

$$l_{1,1} = 2 * \left(\sqrt{r_1^2 - p_1^2} \right) \quad (2.6),$$

Where, p_1 stands for the impact parameter at the beginning of the occultation.

For the second ray, a similar relation gives the following;

$$s_2^2 + p_2^2 = r_2^2 \qquad (2.7)$$

So,

$$s_2 = \sqrt{r_2^2 - p_2^2} \qquad (2.8)$$

Using equations 2.7 and 2.8, we get an expression for the distance traveled by the second ray to be:

$$l_{2,2} = 2 * \left(s_2 - l_{1,2} \right) \qquad (2.9)$$

Generally, for an ionosphere with i layers occultated by j rays, we can generalize the above formulas to get an expression that is most general. Also, as the LEO orbit is assumed to be nearly circular, each r_1, r_2, r_3 and r_4 are equal and let it be r_1. Our final expression (Nava, 2012) then becomes:

$$l_{i,i} = 2 * \left(\sqrt{r_1^2 - p_i^2} - (i-1)l_{1,i} \right) \qquad (2.10)$$

Where, $l_{1,i}$ is the discretization step size of the ionosphere and is equal to a constant value.

The data used in this study is a product of COSMIC mission and is obtained from the COSMIC web site [12]. The file type is in netCDF (network Common Data Form) format with an extension of .nc. A Matlab software is then used to extract the contents of netCDF data one of which is radio occultation $sTEC$ data. The rectangular coordinates obtained from the data helps us to compute the height of the satellite as measured from the center of the earth.

3. Results and discussion

The integrated TEC obtained from COSMIC GPS receivers on 25th October 2015 have been utilized to retrieve the ionospheric electron density profile. The inversion has been carried out by applying the Abel inversion onion peeling algorithm. Each hour RO TEC has been utilized to estimate each hour electron density profile distributions.

The profiles are shown in figures 3.1, 3.2 and 3.3. The occultation TEC is obtained at different locations, because the GPS receiver onboard COSMIC is moving with the satellite. Hence, the profiles shown in those figures are obtained at different locations on the Earth's surface. The variations of the profiles shown can be due to solar zenith variation, spatial variation, and also may be due to some other processes [16].

The peak electron densities vary from 6.6×10^{10} to $1\times10^{12}\,/m^{3}$. Similarly, the peak heights vary from 150 to $300km$. However, for some cases, for example, at 5UT the peak height is found to be about $100km$ and this may be due to spherical symmetry assumption limitations.

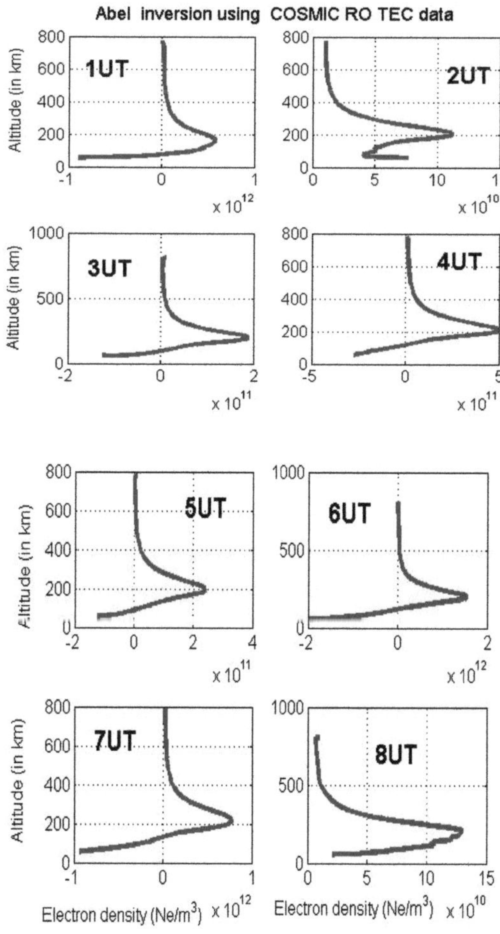

Figure 3.1: Electron density retrieved using Abel inversion from RO COSMIC TEC Data (Peak height < 200km @ 5UT)

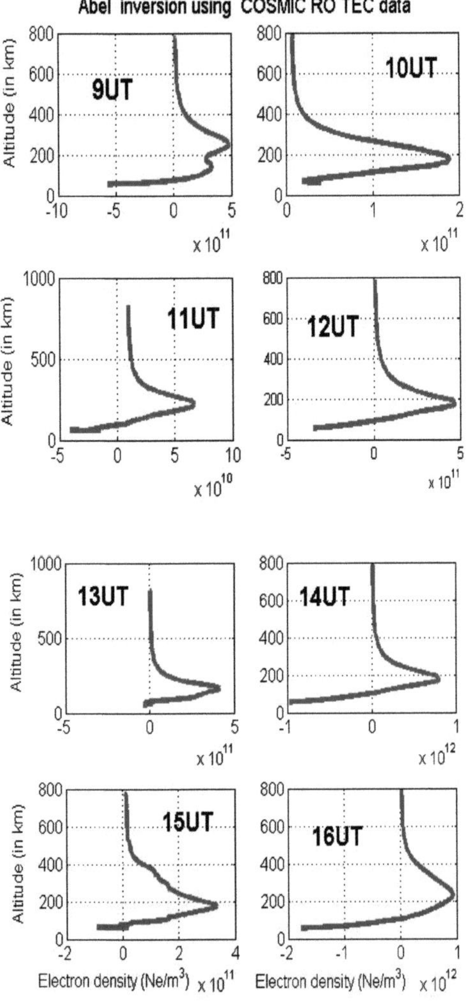

Figure 3.2: Electron density retrieved using Abel inversion from RO COSMIC TEC Data (two peaks @ 9UT)

Observation:

The profiles have shown different altitude variations. In most of the profiles, one peak electron density is shown, but in one case (@ *9UT* and 23UT) two peaks are detected below the F_2 peak (i.e. in the bottom side ionosphere).

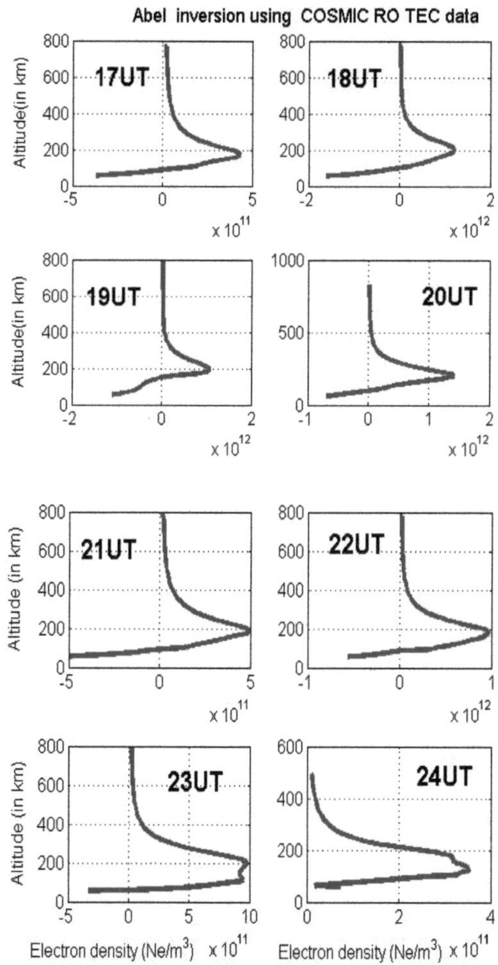

Figure 3.3: Electron density retrieved using Abel inversion from RO COSMIC TEC Data (Peak density @22UT)

These peaks are the peak densities of the F_1 and F_2 layers. Such results without any priori profile information are the benefits of the method utilized in this study. However, in some of the profiles negative electron densities have been obtained in the lower regions of the ionosphere. The negative values of electron density at the lower region are the inherent limitation of the inversion technique implemented. The results are in agreement with the results investigated by Feng

(2010). Spherical symmetry assumptions may account for this unrealistic electron density values (negative electron density values have no meaning and they come simply as a result of the method incorporated which is full of approximations).

In figure 3.4, Abel inversion electron density profiles are superimposed with that of IRI-2007 for the purpose of validating the results of this study. The accuracy of the method used in this study has been evaluated by comparing it against a well-known standard model called IRI. IRI (International Reference Ionosphere) is international standard for ionospheric modeling and parameter estimation [13].

It is based on analytical description of the ionosphere with functions derived from experimental data. IRI, therefore, is said to be an empirical model. The retrieved electron density profiles are plotted up to $2000km$ only to fit the IRI-2007 plots otherwise their values above the height of the COSMIC satellite ($\approx 800km$) are assumed to be zero by the method. As can be seen from the plots, the COSMIC RO retrieved electron density profiles are in agreement with IRI-2007 profiles (both for bottom and topside ionosphere).^

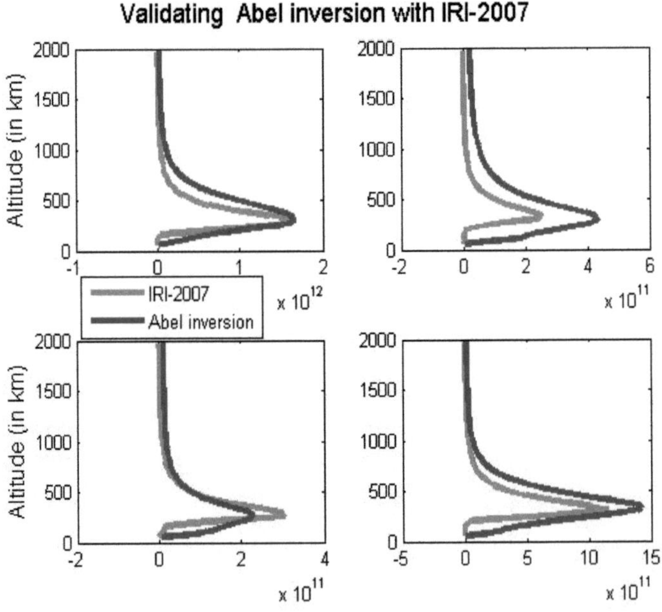

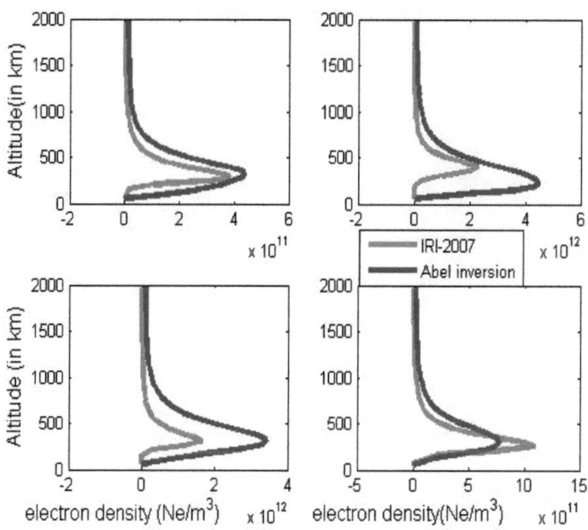

Figure 3.4: Superimposed electron density profiles of Abel inversion (blue) and IRI-2007(red).

However, significant discrepancy between Abel inversion and IRI-2007 peak densities is observed. Unlike the peak densities, the peak heights from these methods show good agreement. Statistical investigations are also carried out to investigate the validity of the Abel inversion results in reference to the outputs of the IRI-2007 standard model
and the results are shown in figure 3.5.

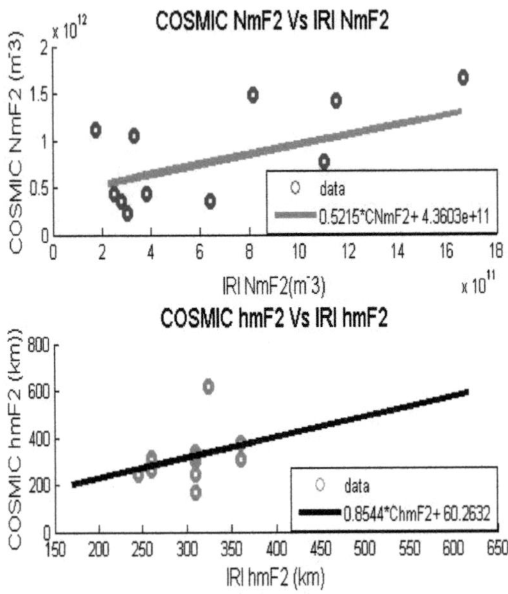

Figure 3.5: The scatter plots of COSMIC NmF2 and hmF2 values against the corresponding IRI-2007 ones.

Figure 3.5 shows scatter plots of NmF2 from Abel inversion versus IRI-2007 (top panel) and hmF2 from Abel inversion versus IRI-2007 (bottom panel). NmF2 and hmF2 are the ionospheric F2 layer peak electron densities and peak heights respectively. As can be seen from the top panel, the correlation coefficient is about 0.5215 indicating that the correlation between Abel inversion and IRI-2007 NmF2 is not that much significant. Hence the two methods show some kind of inconsistency for the case of NmF2 estimation.

The bottom panel, however, shows that the correlation coefficient between hmF2 from Abel inversion and IRI-2007 is 0.8544. hmF2 values from the two methods, therefore, show good correlation and hence Abel inversion and IRI-2007 are found to be consistent in estimating the electron density peak heights and   they can be interchangeably used for the case of hmF2 estimation.

4. Conclusion and recommendation

Conclusion

Ionospheric electron density profiles have been reconstructed from RO TEC data using the Abel inversion onion peeling algorithm. The results are compared with the outputs of IRI-2007 at the same location and time of the RO TEC. From the study carried out in this paper, the following conclusions are drawn.

• The topside and bottom side electron density profiles from Abel inversion and IRI-2007 have shown good agreement; however, they have shown significant difference in their peak electron densities.

• The peak density heights from Abel inversion and IRI-2007 have shown good agreement, which implies the inversion method employed can substitute the standard model IRI-2007 in reproducing the peak heights.

Recommendations

Below are the recommendations that are to be addressed by next work to improve the results of this study.

• Horizontal TEC variations must be taken in to account to overcome the limitations of spherical symmetry assumption used in the electron density retrieval.

• Since electron density retrieval in this study is done up to the COSMIC satellites altitude, it is recommended to carry out a study considering TEC above the LEO orbit height.

Acknowledgements

This work is done by improving the master's thesis work that was done at Bahirdar University during my M.Sc study. My advisor Melessew Nigussie (Ph.D) has played an important role from the very beginning of title selection up to the completion of this paper. Thank you for you had created a good working environment so that I completed the work successfully. Finally, I would like to thank Dr.Endawoke Yizengaw for making the COSMIC TEC data easily accessible.

REFERENCES

[1] Bruno Nava (2012). Data Assimilation into Ionospheric Models. Karl Franzens University Graz, Austria ICTP, June 13, 2012. Trieste, Italy.

[2] http://solar-center.stanford.edu/SID/science/Ionosphere.pdf

[3] Dao Ngoc Hanh Tam (2010). Global Comparisons of NmF2 and hmF2 between COSMIC and Ionosondes. Ms.c thesis, The Graduate Institute of Space Science, Taiwan.

[4] Gary Ouyang, Jian Wang, and Jinling Wang, Generating a 3D TEC Model for Australia with Combined LEO Satellite and ground base GPS Data. School of Surveying and Spatial Information Systems University of New SouthWales, Sydney NSW 2052, Australia, 2008a.

[5] Gary Ouyang, Jinling Wang, Retrieving Vertical Electron Density (VED) Profile with Combined Tomography and Abel Inversion Techniques. School of Surveying and Spatial Information Systems University of New South Wales, Sydney NSW 2052, Australia, 2008b.

[6] Man Feng (2010). Detection of High-Latitude Ionospheric Irregularities from GPS Radio Occultation. Msc thesis, Department of Geomatics Engineering, Uinversity of Calgary.

[7] Mahdi M., Dudy D., and Thoma. H (2013). Ionospheric effects on Microwave Signals. Springer Atmospheric Sciences.

[8] Nigussie, M., Damtie, B., Yizengaw, E., Radicella, S.M., Nava, B. (2013). Validation of the Nequick 2 and IRI-2007 models in East African equatorial region. J.Atmos. Solar-Terr. Phys

[9] Sandro M. Radicella (2010). Ionospheric modeling: Second Workshop on Satellite Navigation Science and Technology for Africa, April 6-24, 2010.Triste, Italy.

[10] S.Skone, ionospheric modeling for wide area diferential GPS applications department of Geomagnetic Engineering, university of Calgary. November 2010.

[11] Tamas I.Gombosi (1998). *Physics of the space environment,* Cambridge University Press, UK.

[12] Xin'an Yue, Bill Schreiner, and Ying-Hwa Kuo (2009). Radio occultation electron density retrieval aided by ground based GNSS observations. COSMIC Program Office, UCAR.

[13] Yizengaw, E., Moldwin, M. B., Dyson, P. L., and Essex, E. A. (2007). Using tomography of GPS TEC to routinely determine ionospheric average electron density profiles. J. Atmos. Solar-Terr. Phys, 69: 314-321.

[14] http://cdaac-www.cosmic.ucar.edu/cdaac/index.html

[15] http://omniweb.qstc.nasa.gov/vitmo/invitmo.html. (accessed 25-oct-2015)

[16] https://www2.bc.edu/endawoke-kassie/ionosphere.html

YOUR KNOWLEDGE HAS VALUE

- We will publish your bachelor's and
  master's thesis, essays and papers

- Your own eBook and book -
  sold worldwide in all relevant shops

- Earn money with each sale

Upload your text at www.GRIN.com
and publish for free